STEAM ON THE UNDERGROUND

The Metropolitan's clumsy mid-1930s style of lettering was not applied to all its locomotives. Nevertheless, 'F' class 0-6-2T No 91 evidently succumbed.
C. R. L. Coles

READ BROTHERS
TLING STORES
15530

STEAM ON THE UNDERGROUND

IAN ALLAN
Publishing

First published 1994

ISBN 0 7110 2282 8

Designed by Ian Allan Studio

Published by Ian Allan Publishing: an imprint of Ian Allan Ltd, Terminal House, Station Approach, Shepperton, Surrey TW17 8AS; and printed by Ian Allan Printing Ltd Coombelands House, Coombelands Lane, Addlestone, Surrey KT15 1HY.

Front cover:
The preserved Metropolitan Railway 'A' class 4-4-0T is pictured restored to Metropolitan Railway livery.
Colour-Rail (LT27)

Back cover, top:
Ex-Great Eastern 2-4-2T No 67200.
Colour-Rail (RRE1010)

Back cover, lower:
Fairburn 2-6-4T No 42231 is seen near Chalfont running over the ex-Metropolitan & Great Central Joint line.
Colour-Rail (BRM421)

Previous page:
Three generations of condensing engines lined up for the photographer at Kentish Town shed on 2 September 1933. From left to right, they are Fowler 2-6-2T No 15530 (later BR No 40031), Johnson 0-4-4T No 1377 (built 1893, retired as BR No 58071 in 1956), and Kirtley 0-4-4WT No 1219 (built 1870, withdrawn 1935).
T. G. Hepburn/Rail Archive Stephenson

Contents

Introduction

Nowadays, on the odd occasions when London's underground railway system is out of action, traffic grinds to a halt and the frustration of motorists is only too evident. It might be thought that traffic jams are a comparatively recent nuisance, but things weren't too different back in the 1850s. In those days, the congestion caused by horsedrawn vehicles was sufficient to generate not only alarm, but also letters to *The Times*. Somewhat disconcertingly, statisticians of the day predicted that if London's traffic continued to increase at a steady rate, the city would by the turn of the century be four feet deep in horse manure. However, that eventuality was avoided due not only to the advent of the motor car, but also to the evolution of underground railways.

For the first 41 years of London's underground railways, steam traction had a near monopoly. The transition from steam to electric traction on London's existing underground network started in July 1905 but, before then, nine different railway companies had operated steam-hauled services on various parts of the system. The spread of electrification obviously curtailed many of the steam workings, but it by no means brought about an early end to those that remained.

Right through to the early 1960s, steam-hauled passenger services still worked to and from Moorgate station via the Widened Lines, and ex-GWR 0-6-0PTs hauled freight trains from the subterranean goods depot at Smithfield meat market. Other regular steam workings of the post-Nationalisation era included the cross-London freight transfers via the Widened Lines, while the route of the old West London Railway through Kensington (Olympia) accommodated a wide variety of steam-hauled freight and excursion trains. At the extremities of what is now the Underground system, steam workings on the former Metropolitan Railway 'main line' beyond Rickmansworth lasted until 1961. Steam traction was also used for passenger workings to Ongar until 1957 and to Chesham until 1960, while High Barnet and Mill Hill East were served by steam-hauled freight workings until 1960. London Transport itself operated a number of steam-hauled departmental workings over electrified lines until June 1971 . . . almost three years after the demise of steam on British Railways.

Although the title of this book is *Steam on the Underground*, some space will nevertheless be devoted to the 'overground' workings on the old West London Railway via Kensington, the ex-LC&DR route via High Holborn and also the enigmatic ex-Metropolitan Railway branch to Brill. The pedantic ones amongst us may quibble about such lines qualifying for inclusion, but the stories of those lines are, it is considered, relevant to the main subject matter. It should be emphasised that this book is not intended to provide a comprehensive history of the various railways featured. The stories of each individual company and line are involved enough to warrant full-length books of their own. Indeed, several have been written.

On a personal note, I vividly recall one of my first trips to London in the 1950s. Travelling on the Underground between Liverpool Street and King's Cross, I was perplexed as to where those 'other' tracks went, particularly as they didn't look like ordinary tube lines. Later on that day, I entered the seemingly deserted Suburban station at King's Cross only for the tranquillity to be shattered by an 'N2' 0-6-2T thrashing out of the dark cavern at the end of the platform. I suspected that the origin of the 'N2's journey was not entirely unconnected with the mysterious 'non-tube'-looking lines I had seen on my Underground trip earlier that day, and I vowed to find out what on earth that engine had been up to in the catacombs under London. It has taken virtually forty years to look into the subject more closely.

Acknowledgements

The biggest thank you must go to Peter Herring, the editor of *Steam Classic* magazine and a born-and-bred Londoner, who gave a wealth of invaluable advice during the preparation of this book.

Many thanks are also due to my wife, Micky, who tells me that her patience is without parallel. I'll not argue.

Greatly-appreciated assistance was also forthcoming from: Richard Casserley, Anne Chovie, Peter Clarke (Buckinghamshire Railway Centre), Rex Kennedy, Roy Miller (Buckinghamshire Railway Centre), Paul Pike, George Rowe, Steve Salmon, John Smith (Lens of Sutton), Graham Stacey (LCGB/RAS), Brian Stephenson, Sheila Taylor (London Transport Museum), Peter Theed, Peter Waller and Ron White (ColourRail).

The 'constant interruption department' was adequately provided by Ozzie (the Spinone pup), Judy (the gem from Bath dogs' home), Pudding (the resident cat) and The Butler (the would-be resident cat).

September 1993

Martin Smith
Coleford, Somerset

Bibliography

Much has previously been written about London's railways and, during the preparation of this book, a degree of secondary-source material has been gleaned from some of the following useful publications:

The Great British Railway Station: King's Cross, Chris Hawkins, Irwell Press.
The History of the GWR, E. T. MacDermot & C. R. Clinker, Ian Allan.
London's Metropolitan Railway, Alan A. Jackson, David & Charles.
Locomotives of the GWR, RCTS.
Locomotives of the LNER, RCTS.
The Metropolitan Railway, C. Baker, Oakwood.
The Metropolitan District Railway, Charles E. Lee, Oakwood.
Railways Through London, C. R. L. Coles, Ian Allan.
Regional History of the Railways of Great Britain; Volume 3, H.P.White, David & Charles.
The Story of London's Underground, John R. Day, London Transport.
The West London Railway, H. V. Borley & R. W. Kidner, Oakwood.
Also *Bradshaw's*, various public and working timetables, and an assortment of contemporary railway periodicals.

The Midland graduated to 0-4-4WTs for its services under London and in 1870 a batch of 20 were built by Dubs & Co to the same design as some earlier 0-4-4WTs. The Dubs-built engines were numbered 780-799 and some survived until the 1930s by which time they had been treated to the luxuries of cabs.

1

The early years

Almost at the very beginning of the railway age, thoughts were given to extending conventional railways into the City of London by means of tunnels, but the major problem with tunnelling was not so much the engineering aspect but money. At the time, any railway company wishing to bore a tunnel was under a legal obligation to purchase the whole of the property under which the tunnel was to pass and, even in the mid-1800s, property in Central London was not cheap. Consequently, the only viable means of building underground railways was either for them to pass beneath streets, thereby minimising the need for purchase of houses, or to promote combined 'road and rail' construction. It was the latter tactic which eventually drew the attention of a City solicitor, Charles Pearson.

In 1851, Pearson joined forces with the civil engineer and architect, John Hargrave Stevens, to formulate a firm proposal for a new road and an underground railway from the Great Northern's terminus at King's Cross to Farringdon Street. The plans made provision for eight parallel tracks into an extensive terminal complex at the southern end of Farringdon Street where there would be facilities for local and main line passenger trains, the handling of freight and the servicing of locomotives. The scheme, known as the City Terminus Railway, was certainly ambitious but it failed to attract financial support and was subsequently put on ice.

The state of play on London's future Underground network by the end of 1884.

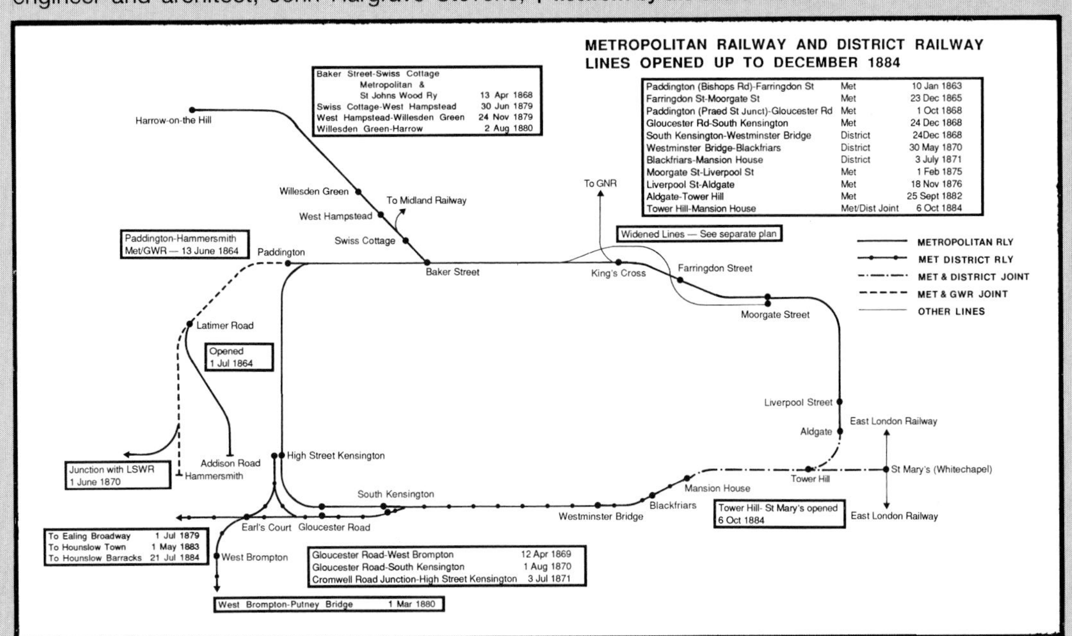

The Metropolitan Railway

The scheme for the City Terminus Railway was resuscitated as an integral part of the proposed Bayswater, Paddington & Holborn Bridge Railway which, on 15 August 1853, was formally incorporated under its revised title of the North Metropolitan Railway. That company was promoted primarily to provide a connection between the Great Western's terminus at Paddington and the City, the GWR itself being painfully aware of the remoteness of its main line terminus. It was proposed that, eastwards from Paddington, the North Metropolitan would pass under New Road, which had been opened in 1756 to become, in effect, London's first bypass. Sections of New Road are better known today as Marylebone Road, Euston Road and Pentonville Road. It had originally been intended that the eastern end of the North Metropolitan would connect with Charles Pearson's City Terminus Railway but, as the Bill for the latter had been thrown out by Parliament, the promoters of the North Metropolitan were presented with a problem. Access to the City, ideally via the City Terminus Railway, was essential if the North Metropolitan's plans were to succeed, but a solution was found by a combination of lateral thinking and sheer desperation.

The North Metropolitan took on board Pearson's grand plans for the City Terminus Railway but subjected them to radical alterations, the proposals for an expansive terminus being abandoned in favour of access to the General Post Office. That was a shrewd move as the revised plans drew the valuable support of the GPO and, predictably, things passed relatively smoothly through the Parliamentary stages. Assent for the railway between Bishop's Road at Paddington and the GPO was gained on 7 August 1854, the name of the railway company being changed once again, this time to the Metropolitan Railway.

Prior to the commencement of construction, various amendments to the plans were made, these including the abandonment of the proposed section from Cowcross Street to the Post Office premises, and a deviation to avoid Clerkenwell Prison. A further amendment was the building of a spur from Copenhagen Cattle Market, just to the north of King's Cross, thereby providing a means of transportation for slaughtered cattle to the new meat market at Smithfield. Because of the anticipated use of the line by the Great Western and Great Northern Railways, it was agreed to lay mixed gauge rails as, of course, the GWR operated on a gauge of 7ft 0in.

Construction of the Metropolitan Railway started in March 1860. In order to accommodate the mixed gauge tracks the tunnels were built to a width of 28ft 6in, although a 25ft width was adopted as standard in later years after the broad gauge workings had ceased. The line was ready for inspection in December 1862 but the Board of Trade inspector, the redoubtable Colonel Yolland, required a few minor improve-

The original terminus of the Metropolitan Railway at Farringdon Street at the time of the opening of the line in 1863. The engine shed, which was owned jointly by the Metropolitan and the GWR, was seldom used by the GWR after the opening of a shed at Hammersmith in June 1864. The shed at Farringdon Street was demolished when the layout was altered during the construction of the Widened Lines in 1865.

FARRINGDON STREET STATION 1863

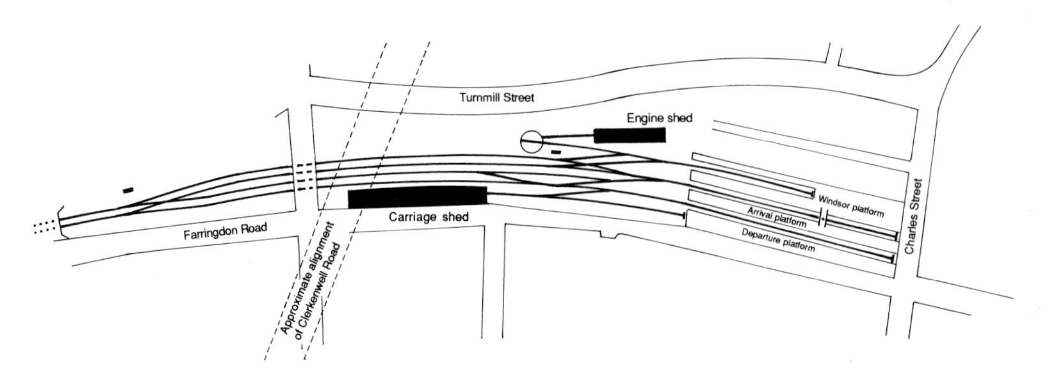

ments to be made before giving the all-clear. A final inspection took place on 3 January 1863 and, over the next five days, a 'rehearsal' of proposed services was undertaken. The serious business of extracting money from fare-paying passengers commenced on Saturday 10 January.

At first, the stations were Bishop's Road (later absorbed into the Paddington complex and renamed accordingly), Edgware Road, Baker Street, Portland Road (renamed Great Portland Street in March 1917), Gower Street (renamed Euston Square in November 1909), King's Cross and Victoria Street (renamed Farringdon Street in May 1863, Farringdon & High Holborn in January 1922 and Farringdon in April 1936). Before very long, the service comprised 67 trains each way on weekdays and 48 on Sundays; the practice of suspending the Sunday services during the hours of church services continued until October 1909. The journey time for the 3¾-mile trip was 18 minutes. From the outset, a significant feature of the Metropolitan's services was the inclusion of workmen's trains, acknowledged as the first such workings in Britain, which had been championed by Charles Pearson. Sadly, Pearson did not see his dreams become reality, as he had passed away on 14 September 1862.

The engineer in charge of the construction work was John (later Sir John) Fowler, and it had fallen to him to organise motive power. In 1860, Fowler had placed an order with Robert Stephenson & Co for a 32ton 2-4-0 tender engine with 5ft 6in coupled wheels and 15in x 24in cylinders. An important requirement of the locomotive's design had been that its exhaust steam should be minimal, and so it had incorporated a high-pressure boiler. The theory had been that the engine could be stoked ferociously when running in open sections, and left to work largely from accumulated pressure when running through tunnels, any exhaust steam being diverted into an injection condenser.

Fowler recorded his theories about his condensing locomotive in his diaries:

'The escape of steam was easily prevented by providing a condenser which is placed under the engine... It consists of a large tank 14ft by 6ft and 2⅔ft deep, containing a large quantity of cold water. When the engine is running, the water is kept in a state of agitation by machinery, and at the same time the waste steam from the cylinder is turned into the tank and is instantly condensed into water. When the water in the condenser becomes warm it is let out of the tank whilst running, if necessary, and fresh cold water from the tender is admitted.

The Condenser is provided with air pumps for creating a partial vacuum and facilitating condensation...

The prevention of the escape of the gases of combustion was, after many designs and experiments had been made, effected in the following manner:

The whole of the fuel necessary for the journey is burnt in the portions of the line which are exposed to the air. This can be done when the engine is running or standing. If the former, the heat which is necessary to raise steam for the time being is used, and the remaining heat stored for use in the tunnel.

The storing of heat was effected partly by raising the steam to an increased pressure (the boiler being made proportionally stronger) but chiefly by filling a large chamber within the boiler with fire clay tubes. Whilst the fuel is being burnt the surplus heat is communicated to the fire clay which absorbs from 1,200 to 1,500 degrees of heat. On entering the tunnel the fire is damped or the blast completely shut off by a contrivance closely resembling tightly fitting venetian blinds... The fire clay then gradually parts with its heat and raises steam in precisely the same manner as the fuel, but as no portion of the fire clay is consumed, no gasses are given out by it as by coke or other fuel.

When the open portion of the railway is reached, the Damper is raised and the fuel immediately fanned bright, reheating the fire clay and preparing the Engine for another subterranean trip.

The engine has run as much as 4½ miles at once without giving out gases or steam.'

In practice, the locomotive's exhaust emission had certainly been kept to a minimum, but so had the contraption's pulling power. Known as 'Fowler's Ghost', the machine seems not to have entered revenue-earning service. Its subsequent story and ultimate fate have, over the years, made it something of a mystery to researchers but, as far as can be determined, it was resuscitated for further trials in 1895 and, it seems, sold to Messrs Beyer Peacock and scrapped.

The Metropolitan's lack of success with its experimental locomotive presented the problem of where the motive power was to come from, but the offer of a working agreement from the GWR was readily accepted by the Metropolitan. It would have been thought that the GWR's celebrated locomotive superintendent, Daniel Gooch, could have designed suitable locomotives in his sleep, but his design for condensing engines proved to be equally unsatisfactory.

Gooch's design was for a class of broad gauge 2-4-0Ts, the condensing pipes of which entered the tanks below the water level. That feature, however, frequently resulted in water being sucked into the cylinders. Water leakage proved a constant problem with the locomotives and, as their tanks were somewhat undersized to start with, Gooch's final design for the GWR did not enter the annals as one of his better attempts. Known as the 'Metropolitan Tanks', 22 of Gooch's 2-4-0Ts were constructed between 1862 and 1864 and, in keeping with broad gauge practice, they carried names instead of numbers. All 22 were withdrawn between 1871 and 1877, but their relatively new boilers were saved and put to stationary use. The boiler from one of the 2-4-0Ts actually finished up on Brunel's famous steamship, the SS *Great Eastern*. The coaching stock used by the GWR on its Metropolitan services consisted of 49 gas-lit eight-wheeled vehicles known as 'Long Charleys', which were considered very spacious and comfortable.

The working agreement between the Metropolitan and the GWR had not been operational for too long before the two companies fell out. The former was quick to criticise the standard of services provided by the latter, while the latter was unimpressed by the tardiness of the former in providing freight facilities. The mighty GWR saw no reason why an upstart such as the Metropolitan should call the tune and abruptly announced that, as from 10 August 1863, it intended to cease providing locomotives and rolling stock.

As a temporary alternative, the Great Northern hastily stepped in with the offer of a working agreement, the company's Superintendent, Archibald Sturrock, having already undertaken preliminary work at Doncaster on smoke-condensing locomotives for the projected GNR workings to Farringdon Street. The GNR supplied a number of 0-6-0s and 0-4-2s for working the Met's services, among them eight '116' class 0-6-0s which had already been earmarked for fitting with condensers. Sturrock's condensing equipment incorporated a pipe and a flexible tube between the chimney and the tender, but, at first, the makeshift joins between the engine and tender sections proved rather too willing to come apart.

The full list of 0-6-0s and 0-4-2s which worked underground in London seems to be unrecorded, but it is nevertheless known that '116' class 0-6-0 No 138 featured among them. In a well-documented episode at Bishop's Road station in May 1864, No 138's boiler exploded, one section being thrown through the roof of Paddington station and damaging the hat of a cab-driver, another landing outside the Dudley Arms in Harrow Road, while the 6cwt dome was found near the Grand Junction Canal, some 200yd away.

Although the Metropolitan's tracks had been laid with mixed gauge rails, the standard gauge ones had never been used and so there was considerable apprehension when the GNR workings commenced. The derailment of an engine at Bishop's Road at 9.30am on the very first day of GNR services did not, fortunately, prove an ill omen, as things subsequently worked reasonably smoothly. The engines supplied by the GNR were far from perfect, but were nevertheless gratefully accepted by the Metropolitan until it took delivery of its own locomotives and rolling stock, the first of the famous Beyer Peacock 4-4-0Ts arriving in June 1864.

In common with the GWR locomotives which had previously been used, the Metropolitan's engines were fitted with condensing apparatus to convert exhaust steam into water which

The mixed gauge tracks of the Metropolitan Railway are clearly distinguishable in this artist's impression of Baker Street station shortly after opening. The official caption for the picture extols the virtues of 'platforms lit partly by natural daylight', but those who experienced early condensing locomotives' inability to condense might have seen another use for the windows.
London Transport

was then returned to the locomotives' tanks. Despite the use of coke instead of coal and the fitting of condensing gear, the locomotives' exhaust still presented enough problems to generate complaints from passengers. At Portland Road station, an extractor fan was installed, while at other stations glass was removed from some windows. Later, ventilation shafts between station ceilings and street level were provided on the Edgware Road-King's Cross section, but unadvertised blasts of steam and smoke above ground did little for the nerves of pedestrians or horses. Contemporary commentators observed that the Metropolitan's regular passengers acquired the habit of growing beards which, if necessary, could be used as air filters. How the engine crews coped with cabs that had only a weatherboard for protection is ill-recorded.

Despite the Metropolitan's apparent disregard for the welfare of its passengers' breathing apparatus, the company successfully appealed to be excluded from the legislation of 1868 which compelled railway companies to include smoking carriages on all trains. It was only the realisation that its competitors provided such facilities that later prompted the Metropolitan to toe the official line.

Despite the friction between the GWR and the Metropolitan, the former did not baulk at using the latter's tracks and inaugurated a short-lived service between Farringdon Street and Windsor on 1 October 1863. On that same date, Great Northern trains started running through to Farringdon Street. Until the opening of a platform on the southbound line at the side of King's Cross main line station, and another on the Hotel Curve for northbound trains in February 1878, the practice was for the GNR's Farringdon Street trains to call at the main line station by means of a reversing manoeuvre.

Usually it was customary for a railway company to provide top-class accommodation for VIPs on official pre-opening trips, but the Metropolitan seemed to think little of carrying several Lords, one Right Honourable and a trio of MPs in open trucks. The date of this inspection trip at Edgware Road station was 24 May 1862.
London Transport

The Metropolitan also extended its original line from Praed Street Junction at Paddington to Gloucester Road on 1 October 1868, and a further extension to South Kensington was opened on 24 December. A junction with the London & South Western at Grove Road, near Hammersmith, was opened on 1 June 1870 to enable GWR trains to work to Richmond although, on 1 October 1877, the Metropolitan inaugurated a service to Richmond via Hammersmith.

Elsewhere, the Metropolitan & St John's Wood Railway opened its line from Baker Street to Swiss Cottage on 13 April 1868 and this was extended to Harrow-on-the-Hill by 2 August 1880. Through workings on the M&SJWR to and from Moorgate Street had, however, ceased as early as March 1869. The semi-independent M&SJWR was worked by its mentor, the Metropolitan Railway, but relations between the two companies were somewhat strained and, at one stage, the Metropolitan denied the M&SJWR access to the 'main line' at Baker Street, thereby forcing through passengers to walk between the two separate platforms. One significant feature of the line northwards from Baker Street was a goods-only connection with the Midland Railway at Finchley Road.

Left:
This is probably the only existing photograph of 'Fowler's Ghost', the experimental condensing locomotive designed and ordered by John Fowler for the Metropolitan Railway. The sheer size and impressive finish of the machine are extremely noteworthy, and so it might be considered that a far more distinguished fate was deserved by this innovative locomotive. The picture is thought to have been taken at Stafford Street bridge, at the eastern end of Edgware Road station, in 1862.
London Transport

Below:
In 1863, the Metropolitan placed an order with Beyer Peacock for 18 standard gauge 4-4-0Ts. Costing £2,600 each and delivered between 20 June and 30 August 1864, they have over the years been attributed with a variety of basic dimensions. Therefore, the dimensions quoted are those given by E.L.Ahrons which, thankfully, tie in closely with what can be determined from official records:

Driving Wheels:	**5ft 9in**
Boiler Pressure:	**130lb**
Leading Wheels:	**3ft 0in**
Water Capacity:	**1,000gal**
Wheelbase (coupled):	**8ft 10in**
(total):	**20ft 9in**
Weights (empty):	**35t 8¾cwt**
(full):	**42t 3cwt**
Maximum axleweight:	**15t 10cwt**
Cylinders (2 outside):	**17in x 24in**
Boiler diameter:	**4ft 0in**
Tractive effort @ 80%:	**13,100lb**
Total heating surfaces:	**1013.8 sqft**

The locomotives were named after Greek Gods. It has been suggested that an art-loving Metropolitan director selected names of Gods who appeared in some of the paintings which were lost when the Titian Gallery at Blenheim Palace was destroyed by fire in 1861. The locomotives' names, which were eventually removed, were:

1 *Jupiter*	**6 *Medusa***	**11 *Latona***	**16 *Achilles***
2 *Mars*	**7 *Orion***	**12 *Cyclops***	**17 *Ixion***
3 *Juno*	**8 *Pluto***	**13 *Daphne***	**18 *Hercules***
4 *Mercury*	**9 *Minerva***	**14 *Dido***	
5 *Apollo*	**10 *Cerberus***	**15 *Aurora***	

This magnificent picture of No 18 *Hercules* in its original condition is believed to have been taken at Hammersmith in the late 1860s.
Ian Allan Library

Above:
The design for the 4-4-0Ts was credited to the Met's engineer, John Fowler. It is often regarded as one of the all-time classics, although Beyer Peacock had, in 1862, built eight similar but smaller locomotives for the Tudela and Bilbao Railway in Spain. An integral part of the Met locomotives' design was, of course, condensing apparatus. This directed exhaust steam from the blastpipe into the tanks by means of a large pipe which had, for part of its length, a smaller pipe inside it to take steam to the lower part of the tanks, thereby aiding circulation.

The Bissel bogie trucks were, perhaps, one of the few unsatisfactory parts of the design; later locomotives had central-pivot bogies of a William Adams design, and those originally built with Bissel trucks were eventually converted. Another niggling fault encountered was that, as the incessant stop-start nature of the locomotives' work placed tremendous strain on the connecting rods, fractures became increasingly commonplace. Therefore, the practice of strengthening the rods with iron was later adopted. Overall, though, the locomotives were very robust, as evidenced by the class leader, No 1, which was not reboilered until 1887, by which time it had notched up 632,145 miles.

By 1870, 26 similar 4-4-0Ts had been purchased by the Metropolitan to complete what was eventually designated the 'A' class. Their numbers were 19-33 and 39-49 but they were all unnamed. This is No 44, one of the 1870 batch.
Bucknall Collection/Ian Allan Library

Right:
A slightly modified version of the 'A' class 4-4-0Ts appeared in 1879, the newcomers having 1,140gal tanks and a total weight of 46tons 15cwt, but smaller boilers and fireboxes. These were designated the 'B' class and 22 were built in all, Nos 34-38 in 1879, Nos 50-59 in 1880, Nos 60-64 in 1884 and Nos 65/66 in 1885. Three of the 'Bs', Nos 57-59, were delivered to the South Eastern Railway but were taken back into Metropolitan stock in 1883. In this picture, 'B' class No 55 of 1880 is seen reversing a Hammersmith train at Aldgate.
LCGB/Ken Nunn Collection

The Hammersmith & City Railway

The GWR and the Metropolitan uneasily buried at least part of the proverbial hatchet to work the Hammersmith & City Railway jointly. The H&CR was incorporated to construct a 2½-mile line from Green Lane Junction at Westbourne Park, about a mile west of Paddington, to Hammersmith, complete with a branch to Kensington (Addison Road). The H&CR's 'main line' opened on 13 June 1864 and the Addison Road branch on 1 July that same year. Addison Road station is, of course, better known these days as Olympia.

Those unfamiliar with the Hammersmith & City's route might have difficulty in co-ordinating the present day intermediate stations with those originally featured. Notting Hill station was eventually renamed Ladbroke Grove, the original Shepherd's Bush station was superseded by a new station on an adjacent site in April 1914, and the original terminus at Hammersmith was replaced by a new station, slightly to the south, in December 1868. Furthermore, Latimer Road station, at the point where the branch diverged, was not opened until December 1868, while Royal Oak was added in October 1871, Wood Lane in May 1908 and Goldhawk Road in April 1914. Of those later stations, Wood Lane remained in permanent use only until November 1914, although it was later used sporadically when special occasions demanded.

The purpose of the Addison Road branch was primarily to provide a connection with the West London Extension Railway, thereby giving a link between Paddington and Battersea which, in later years, sometimes resulted in GWR trains appearing at Victoria, and London, Brighton & South Coast Railway trains materialising at Paddington. The regular local services on the Hammersmith & City line were at first worked by the GWR, mixed gauge tracks being provided from necessity. The working of the line passed to the Metropolitan Railway on 1 April 1865 and the last broad gauge rails between Westbourne Park and Hammersmith were dispensed with as early as March 1869.

Metropolitan 'A' class 4-4-0T No 4 is seen at Hammersmith, shortly after the Met took over the working of the Hammersmith & City line in April 1865. The mixed gauge tracks, which had been used by GWR locomotives since the opening of the line in June 1864, are clearly evident.
London Transport

Above:
The Metropolitan Railway inaugurated its services to Richmond via Hammersmith in October 1877. In this early picture, an unidentified Metropolitan 4-4-0T waits to depart from Richmond with a train for Aldgate.
Lens of Sutton

Below:
For standard gauge passenger workings over the Metropolitan's tracks, the Hammersmith & City and, for a time, the Aldgate to Richmond route, the GWR introduced a class of 2-4-0Ts which, logically, took the name of their broad gauge predecessors, the 'Metropolitan' (or 'Metro') Tanks. The first 20 members of the class materialised between January and May 1869, all being fitted with condensing apparatus, and it appears that the original intention was for the first ten members of each of the next three batches of 'Metros' to have condensing gear. There is conflicting evidence as to whether the master plan was rigidly adhered to but, of the 90 'Metros' built by 1892, around 50 certainly had the necessary fitments. It is virtually impossible to give a definitive list of those which were condenser-fitted as the normal practice was to remove the equipment from locomotives which were transferred out of the London area and to refit it to incomers.

According to official records, the 'Metros' intended to be fitted with condensing apparatus were Nos 455-470, Nos 1096-99 (renumbered 3-6 in 1870), Nos 613-622, Nos 967-976 and Nos 1401-1410. The basic dimensions of the original 'Metros' were: 5ft 0in coupled wheels, 16in x 24in cylinders, 1,080sq ft heating surface, 140lb boilers, 740gal water capacity and weights in working order of 33tons 4cwt; the later 'Metros' were slightly larger, the weights being 33tons 8cwt (Nos 613-22), 34tons 4cwt (Nos 967-976) and 36tons 14cwt (Nos 1401-1410).

The 'Metro' in the picture, No 457, is seen in its original condition and shows off the drop-frames which were not perpetuated by later members of the class. The engine's stay in London was not indefinite, as by the time of the Grouping it was shedded at Ross-on-Wye and, prior to its withdrawal in 1934, it lived at Oxford.
Bucknall Collection/Ian Allan Library

Above:
GWR 'Metro' 2-4-0T No 984 was one of those which had condensing apparatus added at a later date. The framing of this locomotive can be compared to that of No 457, other differences (apart from the later livery) including the fitting of coal rails, a slightly higher weatherboard and toolboxes on the top of the tank.
Bucknall Collection/Ian Allan Library

Below:
The GWR used a system of abbreviations to denote a locomotive's home depot, 'Metro' 2-4-0T No 1407 carrying the relevant plaque on the toolbox on its tank. The letter 'P' denotes Paddington, the depot at Old Oak not opening until 1906, and GWR enthusiasts could argue that this form of 'shedplate' was the forerunner of the system instigated by the LMSR in later years.
Ian Allan Library